A Problem-Based Guide to Basic

Genetics

Donald Cronkite

Hope College

Saunders College Publishing

Harcourt Brace College Publishers

Fort Worth Philadelphia San Diego New York Orlando Austin
San Antonio Toronto Montreal London Sydney Tokyo

Copyright ©1996 by Harcourt Brace & Company

All rights reserved. No part of this publication may be reproduced or transmitted in any form or by any means, electronic or mechanical, including photocopy, recording, or any information storage and retrieval system, without permission in writing from the publisher, except that, until further notice, the contents or parts thereof may be reproduced for instructional purposes by users of BIOLOGY, Fourth Edition by P. Eldra Solomon, Linda R. Berg, Diana W. Martin, and Claude Villee.

Printed in the United States of America.

Cronkite; A Problems-Based Guide to Solving Genetics Problems to accompany Biology, 4E. Solomon.

ISBN 0-03-019089-4

567 095 987654321

SOLVING GENETICS PROBLEMS

Solving genetics problems is not very difficult if you are systematic in your approach. The trick is to recognize the inheritance pattern that each problem contains. Only a few basic patterns exist, so once you have learned to spot those patterns, you will find that you can recognize most any problem you are presented. That's why you need to do many genetics problems to learn genetics. The more problems you do, the more practice you will get at recognizing the patterns.

SEVEN SYSTEMATIC STEPS FOR SOLVING GENETICS PROBLEMS

1. Always use standard designations for the generations and lay out the problem as a simple diagram.

When you cross two organisms, we call those two the **parents** or the **P generation**. Their offspring are called the **first filial generation** or F_1. If you cross two F_1 individuals, their offspring are called the **second filial generation** or F_2. These terms are always used in genetics problems; you should understand the terms and use them whenever working with a problem.

Many problems are stated paragraph form, often covering up the basic pattern with words. You will find it easier to do most problems if you change the information into a simple diagram of the crosses. When you do this, the patterns are much easier to recognize.

Solved Problem 1-1. Although flies usually have a pair of wings, geneticists have found some wingless strains. When one of these wingless flies was crossed to a normal fly with wings, all of their offspring had wings. When some of these offspring were crossed to each other, they produced 428 offspring, of which 320 had normal wings and 108 were wingless.

Summarize this cross in a diagram, using the correct terms for each generation.

Answer The three generations can be summarized in this way:

P: winged × wingless

F_1:   all winged

F_2:   320 winged, 108 wingless

PRACTICE PROBLEMS

1-2. A rare white bison was born in a bison rancher's herd. When it matured, the rancher mated it to a normal bison on three occasions, and each time a normally colored bison was born. When those new bison grew up, two of them were mated to each other. All of their calves were brown except one, which was albino like its grandmother. Summarize this cross in a diagram, using the correct terms for each generation.

1-3. Silkworm blood (or "hemolymph") is either deep yellow or white. The caterpillar may be plain or heavily marked with stripes and spots. A moth with yellow hemolymph that had been a plain caterpillar was crossed to a moth with white hemolymph that had been a marked caterpillar. Among the offspring were 65 yellow marked, 56 yellow plain, 61 white marked and 59 white plain insects. Diagram this cross using the correct terms for each generation.

2. Write down a key for the symbols you are using for the allelic variants of each locus.

In order for genetic crosses to be done, **gene differences** have to be identified. We wouldn't learn much from crossing red-flowered plants generation after generation except that there were no gene differences. An important insight of Mendel's was that the genes act like particles that are passed on from generation to generation, and those particles may occur in more than one alternative form. We summarize this idea by designating each **allele** (the alternative forms of a gene) with a symbol and all the alternative forms of the same gene (a **locus**) with related symbols.

The simplest symbol system designates dominant alleles with an upper case letter and recessive alleles with the same lower case letter. (Specialized notation is used by scientists studying particular organisms, such as Drosophila. Simple notation is sufficient for our purposes, however.) For example, in humans the recessive allele for sickle cell anemia is symbolized by s, and the allele for the normal hemoglobin is S. The symbols only tell us which allele is dominant and which recessive, not which one is "normal" or the most frequent. For example, the dominant mutation "Bar eye" in fruit flies reduces the number of facets in the compound eye. Most flies with normally shaped eyes have the recessive allele of this gene. So Bar eye is B and normal eye is b.

When working with diploid organisms, each individual will contain two alleles for each locus. If the individual is **homozygous**, we symbolize it with two of the same allele symbol, but if **heterozygous**, we use two different allele symbols. Gametes are designated with single allele symbols for each locus. If more than one locus is being symbolized in one individual, separate the symbols of different loci with a space.

Solved Problem 2-1. Mendel studied seven different loci in peas. Each locus has two possible alleles. Let tall be T and short be t and red flower be R and white be r. Then write the symbols for each of the following:

A) a homozygous tall plant
B) a heterozygous red plant
C) a short plant with white flowers
D) a plant heterozygous at both loci
E) a gamete with the tall allele
F) a gamete with the short and red alleles

Answer A) TT   B) Rr   C) tt rr   D) Tt Rr   E) T   F) t r

Solved Problem 2-2. Drosophila usually have round eyes, but a recessive mutation results in eyeless flies. They usually have lovely straight bristles on their backs, but a recessive mutation results in crooked bristles that are twisted like little corkscrews. Finally, a dominant mutation changes the normal brick red eyes to a purple color called "plum." Symbolize a fly heterozygous at all three loci.

Answer You will need to decide what letter you want to use for each locus. Suppose you use e for eyeless, c for crooked and P for plum. Then the fly in question would be Ee Cc Pp.

PRACTICE PROBLEMS

2-3. Huntington's disease is a human inherited disorder caused by a deleterious alteration (mutation) of a normal allele at the Huntington's gene locus. This mutant allele is dominant over the normal allele, and is not expressed until mid-life. Cystic fibrosis is a human recessive disorder caused by a mutation in both alleles of the cystic fibrosis gene. Choose what letters you want to use and then write the symbols for each of the following individuals.

A) A young person who is destined to have Huntington's disease later in life.

B) A person who has neither Huntington's disease nor cystic fibrosis.

C) A person who has cystic fibrosis but will not develop Huntington's disease.

2-4. A homozygous fly with an ebony body color was crossed to a fly homozygous for the wild type gray body, and all their offspring were gray. Write symbols for each parent and their offspring.

3. Determine the genotypes of the parents of each cross.

Geneticists deal with two general features of organisms. The **genotype** is the actual collection of genes that an organism has, while the **phenotype** is a description of the characteristics of an organism. Most simple genetics problems are of two general kinds. One, which will be covered in this section, requires you to find the genotypes or phenotypes of parents, while the other, covered in sections 4-7, requires you to find the genotypes or phenotypes of offspring. Three kinds of evidence can be used to figure out the parents.

A. Are they from true-breeding lines?

If a line results in the same kind of offspring, generation after generation, we say they are **true breeding**. Such a line must be homozygous. Sometimes a problem will simply say that the lines used are true breeding. Sometimes other data allow you to figure that out.

Solved Problem 3-1. Which of the following individuals are probably homozygous?

A. A rat from a true breeding strain of laboratory animals.

B. Two parents with brown eyes who produce a child with blue eyes.

C. A snapdragon with purple flowers from a strain that had consistently produced purple flowers for 14 generations was chosen for a cross.

Answer
A. If the rat is from a true breeding strain, that means it is, by definition, homozygous.

B. The parents are not true breeding, so at least one of them is not homozygous, and probably neither is.

C. "True breeding" means that the strain produces the same kind of individuals in every generation, so this plant is from a true breeding strain and thus is homozygous.

Practice Problem

3-2. Jake made a deal with his father-in-law Lavern. They divided up Lavern's flock so that Jake got all the spotted goats and the black sheep while Lavern got the solid colored goats and the white sheep. They agreed that whenever a spotted goat or a black sheep appeared in Lavern's flock, he would give it to Jake, and whenever a solid goat or a white sheep appeared in Jake's flock, he would give it to Lavern. As goats and sheep were born in the flocks, Lavern had to keep giving some of the offspring in his flock to Jake, but Jake didn't have to give any of his flock to Lavern. Explain this, using the terminology of this section.

B. Can their genotypes be reliably deduced from their phenotypes?

A **dominant** allele is one that is expressed whether homozygous or heterozygous, while a **recessive** allele is only expressed when homozygous. If we know which allele is dominant and which recessive, we can know the genotypes of recessive parents with certainty. If we know nothing about an individual except that it has the dominant phenotype, then its genotype will be ambiguous — it could be either homozygous or heterozygous, and other information will be needed to know which it is for certain.

Solved Problem 3-3. Look at the cross described in Problem 1-1. Suppose that lack of wings is recessive and is symbolized with a lower case w and the allele for having wings is W. On the basis of that information alone, write the genotypes of all the individuals in the parental, F_1 and F_2 generations that you know for certain.

Answer All the wingless flies are unambiguously ww. On the basis of only our knowledge of the parental phenotypes, we are not able to tell with certainty what the genotypes of winged flies might be.

P: winged   ×   wingless
           ?   ww

F_1:   all winged
                     ?

F_2 320 winged, 108 wingless
                ?   ww

Solved Problem 3-4. Smooth fox terriers sometimes have congenital myasthenia gravis, a muscle disease due to a recessive allele that causes a defect in the nerve muscle synapse when homozygous. Such dogs almost always die by 6-9 weeks of age. Can we ever know for certain the genotype of prospective parent fox terriers by looking at their phenotypes?

Answer No, we cannot. The recessive homozygotes have myasthenia gravis and die as puppies. So they can never be parents. Prospective parents do not have myasthenia gravis, so they could be homozygous dominant or heterozygous.

Practice Problems

3-5. Refer to problem 3-2. Assuming that these coat colors are due to single gene differences, can we know any of the genotypes with certainty?

C. Do the phenotypes of their offspring provide any information?

A diploid organism provides one allele of each locus to its offspring. The offspring receives one allele from each parent. Sometimes this makes it possible to deduce the genotype of a parent by looking at the offspring. An organism that has a recessive phenotype must be homozygous recessive and must have received one recessive allele from each parent. So each parent must have at least one recessive allele to give to the offspring. If a parent has the dominant phenotype and there are offspring with the recessive phenotype, that parent must be heterozygous.

Solved Problem 3-6. A man's wife had an albino child. Albinism, a lack of melanin pigment, is due to a recessive allele of a single gene. "Neither you nor I is albino," said the man. "Therefore, I am not the father of this child." Is the man correct?

Answer Let's symbolize the recessive albino allele as a and the dominant allele that produces melanin as A. Let's diagram the cross and write what we know for sure.

P: melanin   ×   melanin
       ?   ?

F_1 one albino child
              aa   [We know the genotype because the phenotype is the recessive one.]

Because the parents are not albino, they could be either AA or Aa. A way of symbolizing that is to write their genotypes as A_, where the blank could be either A or a.

P: melanin   ×   melanin
     A_   A_

F_1 one albino child
              aa

The albino child must have received an albino allele from each parent in order to be aa, so that allows us to fill in the blanks of the parents genotypes.

P: melanin   ×   melanin
      Aa   Aa

F_1 one albino child
              aa

So the man could be the father of the child even if he himself nor any of his relatives is albino.

Solved Problem 3-7. Refer to Problem 3-2 and 3-5. If Lavern knew anything about genetics, he might prefer to kill and eat the ewes (female sheep) in his flock that gave birth to black sheep. Would that be a good strategy? Why?

Answer White ewes have the dominant phenotype. If they give birth to black sheep (ww), they must be Ww, since the offspring get a w from each parent. If he eliminates the heterozygotes, he won't have to keep giving his sheep away to his son-in-law.

PRACTICE PROBLEMS

3-8. Refer to Problem 3-3. What can we know for certain about the genotypes of the winged flies?

3-9. Farmer O'Sullivan has a prize boar named Honey which has won many honors at the State Fair. Several other farmers want to mate Honey to their sows so they can have prize pigs too. But Farmer O'Sullivan knows that some of Honey's litter mates have been runts and that the failure to grow properly was due to a recessive gene. What does this tell us about Honey's possible genotypes?

3-10. What cross could Farmer O'Sullivan do to find out for certain what Honey's genotype is?

4. Indicate the possible kinds of gametes formed by each of the parents.

Sexual reproduction involves the alternation of **fertilization**, which produces zygotes, and **meiosis**, which produces gametes. Your thinking about genetic problems should center on that alternation of fertilization — gamete formation — fertilization — gamete formation in laying out the problem. Then, by systematically filling out what we know and using that information to deduce what we don't know, we can solve almost any genetics problem. In this section we will work out the gamete types. In order to keep the operation systematic, we will circle the gametes that are produced to distinguish them from the diploid genotypes that produce them.

A. If it is a monohybrid cross, we must apply the Principle of Segregation.

In a **monohybrid cross** we focus on a single locus. When there are just two possible alleles, gamete formation is simple. Each gamete will receive one allele. If the parent is homozygous, all its gametes will be alike. If the parent is heterozygous, each gamete will receive only one of the two possible alleles. This is what is meant by the **Principle of Segregation.**

Solved Problem 4-1. What are all the types of gametes formed by individuals of each of the following kinds? What proportion of every kind of gamete will each individual produce?

A. QQ   B. Ww   C. mm

Answer Remember, each gamete will receive only one allele, and if there are two possible alleles, half the gametes will get each type.

A. All the gametes produced by QQ will be Q.

B. Half the gametes produced by Ww will be W and half will be w.

C. All the gametes produced by mm will be m.

PRACTICE PROBLEMS

4-2. What are all the kinds of gametes that chickens from a true breeding white-feathered strain can produce?

4-3. Refer to Problem 3-6. What are all the kinds of gametes that each parent in that problem can produce?

B. If it is a dihybrid cross, we must apply both the Principle of Segregation and the Principle of Independent Assortment.

If two loci are being studied simultaneously (a **dihybrid cross**), then the problem might seem a little more complicated, but it really isn't. If you have two loci, you can treat each like a monohybrid cross. Even if there are many more loci than two, you can still do this and have a simpler problem to solve. Each gamete will only get one allele from each locus. In other words, all the loci are acting according to the Principle of Segregation.

Every locus will have an allele represented in every gamete. A particular allele of one locus can be combined with either of the alleles of the other locus. The chance that an allele of one locus will end up with an allele of the other locus depends only on the frequency with which each allele is found in gametes. That is what is meant by the **Principle of Independent Assortment.**

The important thing is to be sure that you have accounted for all of the possible combinations of alleles when you figure out the gametes. Two accounting methods make this rather easy.

1. The branch method. Write down the phenotype of the first locus, and then make a little branch showing the kinds of gametes expected from that locus. Suppose, for example, that we were working with an organism heterozygous for two loci — Aa Bb.

Write a branch diagram for the first locus like this.

The A allele and the a allele can now each be combined with either of the choices for the next locus, so write a second branch diagram next to the end of the first branches. Circle the resulting gametes.

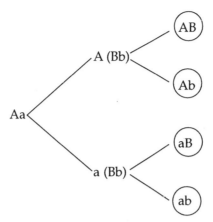

2. The alternating list method. If you know how many gametes you should end up with, then it is easier to be sure you have listed all the possible types. This second method for accounting for all the gametes is based on the observation that the number of different kinds of gametes will be equal to 2^n where n is the number of loci that are heterozygous. Again suppose the organism is the double heterozygote <u>Aa</u> <u>Bb</u>. Because there are two heterozygous loci, there will be $2^2 = 4$ different kinds of gametes. Of the four gametes, two will have an <u>A</u> allele and two will have an <u>a</u> allele. List these like this:

    A

    A

    a

    a

Of the gametes with an <u>A</u> allele, half will have <u>B</u> and half will have <u>b</u>. Likewise, half of those with the <u>a</u> allele will have <u>B</u> and half will have <u>b</u>. So now write a second column on your list like this:

The alleles of the last locus should alternate every other one as in this case if you did it right. Since the list is complete, we've circled them to indicate that they are gametes.

Although we will be concentrating on crosses with only two loci, both the branch and the list methods will work no matter how many loci you are working with. If you want to be sure you have the idea for the dihybrid cases, try some of the later problems in this section to see if you can do it with a larger number of loci.

Solved Problem 4-4. In each of the following cases use either the branch or the list method to determine the kinds and proportions of gametes that the individual will produce.

    A. Mm Nn   B. AA bb   C. RR Tt

Answer. A. This is exactly like the example except that M and N have been substituted for A and B. The four possible gametes are MN, Mn, mN and mn.
B. Remember that the two loci act independently. In this case the first locus is homozygous and can only produce gametes containing an A allele. Likewise, because the second locus is homozygous, it too can produce only one kind of gamete, which contains allele b. So the only kind of gamete possible is Ab. Incidentally, if you use the alternating list method, n = 0 since there are no heterozygous alleles, and the number of different gametes is $2^0 = 1$.
C. To do this one by the branch method, first consider the locus that is RR. It will make only one kind of gamete, containing R.

              RR   R

Then the second locus can form either T or t gametes, so

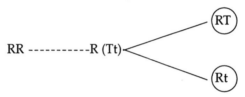

By the list method, first determine the number of possible gamete types, which is $2^1 = 2$ since there is only one heterozygous locus. There will be two gametes, and all will contain R.

              R

              R

Then half of those will contain T and half t, so

Solved Problem 4-6. Tay-Sachs disease is a serious human recessive disorder which affects newborn infants who are homozygous recessive for the deleterious allele t. People with a single normal allele t do not have Tay-Sachs disease. However they can be detected as heterozygotes for this locus by biochemical screening. A dominant gene P produces a characteristic called "polydactyly" in which humans have one or more extra fingers or toes. A person with polydactyly and one normal parent is found to be heterozygous for Tay-Sachs disease. What are all the kinds of gametes this person can produce?

Answer We know the person is heterozygous Tt because the problem says so. A person with the dominant trait polydactyly is either PP or Pp. Because the person has one normal parent (pp), that parent would have produced only p gametes, so the person under consideration must be pp. The question thus breaks down to "What are all the kinds of gametes a person can produce who is Pp Tt?" The answer is PT, Pt, pT, and pt.

Solved Problem 4-7. Write all the kinds of gametes that can be produced by a person with this genotype: Aa BB cc Dd Ee ff gg HH.

Answer At first this looks rather daunting, but take it one step at a time. Let's use the list method, although either method would work. There are 8 loci in this problem, but only three are heterozygous. So the total number of different gametes that can be produced is $2^3 = 8$. Half will contain allele $\underline{A}$ and half will contain allele $\underline{a}$. So start the alternating list this way:

>A
>A
>A
>A
>a
>a
>a
>a

Now all those gametes will have a $\underline{B}$ allele and a $\underline{c}$ allele because these two loci are homozygous.

>ABc
>ABc
>ABc
>ABc
>aBc
>aBc
>aBc
>aBc

The next two loci are heterozygous, and in each case an upper case allele will end up in half the gametes and a lower case allele in the other half. Do the alternating list for each.

>ABcDE
>ABcDe
>ABcdE
>ABcde
>aBcDE
>aBcDe
>aBcdE
>aBcde

The last three alleles are homozygous, so the same allele will end up in each gamete. The final list will therefore look like this:

>ABcDEfgH
>ABcDefgH
>ABcdEfgH
>ABcdefgH
>aBcDEfgH
>aBcDefgH
>aBcdEfgH
>aBcdefgH

PRACTICE PROBLEMS

4-8. Farmer Vandermeer had two hogs named Olga and Hank who were normal healthy hogs, but whenever he crossed them, some of their offspring had cleft palates. Assuming cleft palate has a simple genetic basis, what are all the kinds of gametes that Olga and Hank can produce?

4-9. If a Drosophila is heterozygous for the recessive alleles e (ebony), f (forked), t (temperature-sensitive) and v (vestigial), what are all the kinds of gametes that fly can produce?

5. Set up a Punnett Square.

Fertilization follows gamete production, and at fertilization we want to combine the gametes of one parent in all the possible combinations with the gametes of the other parent. A simple way to do that is to construct a **Punnett Square**. This is a grid with the gametes of one parent written across the top and the gametes of the other written along the left side.

Solved Problem 5-1. Suppose we crossed two organisms each of which had the genotype Bb. Set up a Punnett Square for that cross. What if one parent were Bb and the other were bb?

Answer Set out a simple diagram of the cross to show what kinds of gametes each parent will produce.

P: Bb × Bb

Gametes: (B) and (b)   (B) and (b)

Then construct the Punnett Square.

	B	b
B		
b		

If the parents were Bb and bb, the gametes will be a little different, but the approach will be the same.

P: Bb × bb

Gametes: (B) and (b)   all (b)

	B	b
b		

Solved Problem 5-2. Set up the Punnett Square for the cross of parents that are both <u>Aa</u> <u>Bb</u>.

Answer Again do a simple diagram of the cross and then use the resulting gametes to set up the Punnett Square.

P: Aa Bb   ×   Aa Bb

Gametes: (AB) (Ab) (aB) (ab)   (AB) (Ab) (aB) (ab)

	AB	Ab	aB	ab
AB				
Ab				
aB				
ab				

PRACTICE PROBLEMS

5-3. Suppose we cross an organism of genotype <u>MM</u> <u>NN</u> with one that is <u>mm</u> <u>nn</u>. Prepare a Punnett square for this cross.

5-4. Set up a Punnett Square for two flies heterozygous for <u>B</u>, <u>F</u> and <u>G</u>.

6. Fill in the Punnett Square and determine the genotypic and phenotypic ratios of the offspring.

With the Punnett square set up, it is now a simple matter to fill in all the intersections with the alleles from both gametes that form the intersection. In this way we can be sure of having produced all possible gamete combinations. Then, once the genotypes of the new zygotes are determined, you can figure out the phenotypes of each zygote type to complete the problem.

Solved Problem 6-1. Fill in all the intersections on the Punnett Squares of Problems 5-1 and 5-2.

Answer At each intersection of the square, fill in all of the gametes from the row and column that intersect.

5-1

	B	b
B	BB	Bb
b	Bb	bb

	B	b
b	Bb	bb

5-2

	AB	Ab	aB	ab
AB	AABB	AABb	AaBB	AaBb
Ab	AABb	AAbb	AaBb	Aabb
aB	AaBB	AaBb	aaBB	aaBb
ab	AaBb	Aabb	aaBb	aabb

When Punnett Squares get a little complicated as in 5-2, follow a systematic pattern for filling in the intersections so you won't get confused.
1) Fill in just one allele at a time, and keep the alleles of the same locus together. So in the example above, not that all the A's are together in any one intersection and all the B's are together. Fill in the A's first. Then go back for the B's.
2) When there are two different alleles in an intersection, put the upper case letter first. Notice that there are no genotypes written aA or bB.

PRACTICE PROBLEMS

6-2. Fill in all the intersections on the Punnett Square of Problem 5-3.
6-3. Fill in all the intersections on the Punnett Square of Problem 5-4.

7. If you do not need to know the frequencies of all of the expected genotypes and phenotypes, you may use the rules of probability as a shortcut.

Mendel's Law of Independent Assortment tells us that the probability of an allele from one locus going into a gamete is independent of the probability of an allele from another locus going into that gamete. Assorting alleles into gametes are independent events. There is a rule of probability called the **Product Rule** which says that you can find the probability that two independent events will happen at the same time by multiplying their two independent probabilities. When loci are heterozygous (say Aa), the probability that a gamete will receive the recessive allele a is 1/2, and the probability of receiving the dominant allele A is also 1/2. So what is the probability of a homozygous AA? It is $1/2 \times 1/2 = 1/4$.

Then take this one step further. If the parents are heterozygous at two loci (Aa Bb), the probability of getting an offspring that is homozygous for both A and B is $1/4 \times 1/4 = 1/16$.

Another rule of probability, the **Sum Rule** deals with mutually exclusive events. If an event can happen either one way or another, we say the events are mutually exclusive. For example, if we wonder how an Aa individual could be produced by two Aa parents, we can see that there are two ways that could happen: the first parent could contribute an A and the second an a. Or the first parent could contribute an a and the second an A. If the Aa is produced one way, it won't be produced the other, so these are mutally exclusive events. If we want to know the probability that something could occur by several different mutually exclusive methods, we add (or "sum") the probabilities of each method. So in the case of the two Aa parents producing an Aa offspring, the chance of doing this by getting A from the first parent is 1/4, and the chance of doing this by getting a from the first parent is 1/4, so the chance of getting Aa offspring is $1/4 + 1/4 = 1/2$.

Solved Problem 7-1. A man is heterozygous for an allele for sickle cell anemia, which is expressed only when homozygous. His wife is also heterozygous for sickle cell anemia. Assuming that male and female children are each born about half the time, what is the probability that a first child will be a son with sickle cell anemia? What is the probability that their first child will be a normal daughter?

Answer If both parents are heterozygous for the sickle cell allele, the probability of having a child with sickle cell anemia is 1/4, and the probability of having a child without the disease is 3/4. The probability of having a son is 1/2, so the chance of having a son with sickle cell anemia is $1/4 \times 1/2 = 1/8$. The chance of having a daughter without sickle cell anemia is $3/4 \times 1/2 = 3/8$.

Solved Problem 7-2. In cattle red and white coat color is controlled by a recessive allele r; the dominant allele R causes black and white coat color. Cattle can either lack horns (HH or Hh) or have them (hh). If we cross two double heterozygotes, what will be the proportion of each kind of offspring?

Answer Consider each locus separately. A cross of Rr × Rr will produce 1/4 red and white and 3/4 black and white. Hh × Hh will produce 3/4 hornless and 1/4 horned. To predict the next generation, therefore, we can simply multiply probabilities:

Red and white, horned $1/4 \times 1/4 = 1/16$
Red and white, hornless $1/4 \times 3/4 = 3/16$
Black and white, horned $3/4 \times 1/4 = 3/16$
Black and white, hornless $3/4 \times 3/4 = 9/16$

PRACTICE PROBLEMS

7-3. Suppose there was a cross in which each parent was heterozygous for 10 different loci. What proportion of the offspring will be homozygous for all 10 recessive alleles?

7-4. Refer to problem 7-2. A farmer said to a geneticist, "If I cross these 15 heterozygous black and white horned cows to a red and white bull that I'm pretty sure is heterozygous for the hornless gene, and I get two calves from each mating, how many of the calves can I expect to be red and white hornless cows, and how many can I expect to look like their father?

GOING BEYOND THE SEVEN STEPS

8. The combination of the seven steps that you use depends on the kind of problem you have to solve.

In general there are two kinds of genetics problems, those that tell you about the parents and ask for conclusions about the offspring and those that tell you about the offspring and ask for conclusions about the parents. These problems require slightly different approaches.

A. Given parental data, make conclusions about the offspring.

This is the most direct kind of problem. Steps 4-7 are most helpful, although you will still benefit from simplifying the problem with a diagram, and you will need to define the gene symbols to be used.

Solved Problem 8-1. The p allele of the German cockroach produces a fine mottling all over the body. This condition is called "peppery" and is recessive to the wild type allele lacing mottling ("normal" coloration). What will be the genotype and phenotype ratios among the offspring in each of these crosses: A) a cross between two peppery cockroaches. B) A cross between two heterozygous cockroaches. C) a cross between a peppery cockroach and a heterozygote cockroach.

Answer. Diagram each cross, put down all that you know, and then construct Punnett Squares to get the answer.

A) P: Peppery × Peppery
        pp   pp

   Gametes: all Ⓟ   all Ⓟ

   Punnett Square:

                    p
                p | pp |

All the offspring are peppery

B) P: Normal × Normal
        Pp   Pp

   Gametes: 1/2 Ⓟ; 1/2 Ⓟ   1/2 Ⓟ; 1/2 Ⓟ

   Punnett Square:
                        P   p
                P | PP Normal | Pp Normal |
                p | Pp Normal | pp peppery |

3/4 normal and 1/4 peppery

C) P: Peppery × Normal (heterozygote)
        pp   Pp

   Gametes: all Ⓟ   1/2 Ⓟ; 1/2 Ⓟ

   Punnett Square:
                        P   p
                p | Pp Normal | pp peppery |

1/2 normal and 1/2 peppery

Solved Problem 8-2. A plant had the genotype Ff Gg Hh. Capital letters indicate dominant alleles. If two such plants were crossed, A) what proportion would show the dominant phenotype for F? B) What proportion would show the recessive phenotype for all three loci?

Answer Since there are three heterozygous loci, there will be 8 different possible gametes:

FGH	fGH
FGh	fGg
FgH	fgH
Fgh	fgh

If you do these problems with a Punnett Square, there will be 8 × 8 = 64 boxes to fill in, but you don't have to do that, given what the problem asks. You can use the multiplication rule of probability.

A) This question is only about F and f. The loci are independently assorting, so you don't have to consider G and H. In a cross Ff × Ff, 1/4 will be FF and 1/2 will be Ff. So 3/4 will have the dominant phenotype.

B) In a cross of Ff X Ff, the ff will occur 1/4 of the time. Likewise, in a cross of Gg × Gg, the gg will occur 1/4 of the time, and in a cross of Hh × Hh, the hh will occur 1/4 of the time. So the genotype ff gg hh will occur 1/4 × 1/4 × 1/4 = 1/64 of the time.

PRACTICE PROBLEMS

8-3. Look at Problem 8-2. If one of the triple heterozygotes were crossed to a plant homozygous for all three recessive alleles, what would be the proportions of each genotype among the offspring?

8-4. In humans, polydactyly is due to a dominant allele and results in extra fingers and/or toes. Phenylketonuria is recessive and is a condition due to a disorder in the metabolism of phenylalanine. Unless given a special diet as infants, people with phenylketonuria may have varying levels of mental retardation. A man who has neither condition but whose father has phenylketonuria and a woman with polydactyly like her father but without the phenylketonuria allele wonder what the probability is of having a child with both conditions. What about with one of these conditions?

B. Given data about progeny, make conclusions about parents.

Here you will work backward. After you have done a number of problems of the first kind, you will see that there are just a few possible ratios that you get among the offspring. Recognizing these ratios is very important, because when you know the offspring ratios, that is almost always useful in making conclusions about the parents.

If the offspring ratio is	Then we know this about the parents	The cross is called
3:1	Heterozygous at one locus. One allele is dominant.	Monohybrid
1:1	One parent homozygous recessive. One parent heterozygous.	Test cross
9:3:3:1	Heterozygous at two loci. A dominant allele at each locus.	Dihybrid
1:1:1:1	One parent a double heterozygote. One parent homozygous recessive for both loci.	Two point test cross

Solved Problem 8-5. A tropical fish fancier produced an albino zebra fish. It lacked the blue stripes characteristic of the normal zebra fish, and it had pink eyes. When he crossed a normal zebra fish to the albino, all the offspring were normal. When he crossed two of the normal F_1 offspring, 1/4 were albino and 3/4 were normal. When an albino was crossed to the normal F_1, half the progeny were albino and half were normal. What is the genotype of each fish mentioned in this problem?

Answer Look for a cross that gives readily familiar results. There is a three to one ratio among the progeny of the F_1. This is what we expect when two individuals are crossed that are both heterozygous at a single locus. So that cross would be:

Normal × Normal
Aa   Aa

When albino is crossed to normal, the offspring are normal, suggesting that albino is the recessive character. So the original parental fish were a normal AA and an albino aa.

The last cross of albino with the F_1 normal confirms these previous identifications. The 1:1 ratio in that cross is characteristic of a test cross in which one fish (the normal) is Aa and the other fish (the albino) is aa.

Solved Problem 8-6. A mutation in chickens resulted in extra bones in the wings and feet, a shortened beak, and an inability to hatch because they are unable to peck their way out of the shell. The mutation is a recessive allele called d for "diplopodia." A cross was done involving diplopodia and another locus that controls feather color called "dominant white." The I allele results in white feathers, but I results in colored feathers.

In a particular cross the offspring included 180 white chickens, 68 with colored feathers and diplopodia, 166 with colored feathers and 53 with white feathers and diplopodia. Figure out the genotypes and phenotypes of the parents.

Answer Since we know the dominant relations of these genes, we can write down quite a bit that is known about the offspring without considering the ratios at all.

180 white (no diplopodia)
68 color and diplopodia
166 color (no diplopodia)
53 white and diplopodia

White is dominant to color and normal is dominant to diplopodia, so:

180 white (no diplopodia)   I _ D _
68 color and diplopodia   ii dd
166 color (no diplopodia)   ii D _
53 white and diplopodia   I _ dd

Now consider each locus separately to simplify the problem. (Notice how often simplification comes from the approach of considering each locus separately.)

The feather color phenotypes occur in a ratio of about 1:1 (233 white: 234 color). That is characteristic of a test cross Ii × ii.

The diplopodia phenotypes occur in a ratio of about 3:1 (346:121). That is characteristic of a cross of two heterozygotes where one allele is dominant (Dd × Dd).

So the original cross was probably Ii Dd × ii Dd.

PRACTICE PROBLEMS

8-7. A cross in chickens involved the diplopodia locus (See Problem 8-6) and "naked neck" (n), a recessive mutation at another locus that results in loss of feathers on the neck. The offspring included 81 normal, 4 with naked neck and diplopodia, 22 with naked neck and 34 with diplopodia. What were the genotypes and phenotypes of the parents?

8-8. Horses can have big white spots (a recessive b) or little white spots (a dominant B). They can also have weak hooves (W) or strong hooves (w) and normal nostrils (F) or flaring nostrils (f). A woman admired a horse named Izzy and wanted to buy a colt that had Izzy as a parent. Izzy's owner wanted to sell her Lucinda, a cute colt with big white spots, weak hooves and flaring nostrils. Izzy has little white spots, strong hooves and normal nostrils. Could Izzy be a parent of Lucinda?

9. Many genetics problems go beyond these simplest situations.

A. Complex relations between genotype and phenotype.

1. Incomplete dominance.

When there is **dominance**, the heterozygote has the same phenotype as the dominant homozygote. However, dominance is not always observed. Many genes show **incomplete dominance** in which the heterozygote has a phenotype intermediate between the two homozygotes. For example, a heterozygote with alleles for red flowers and for white flowers might be pink. Incomplete dominance makes your life a little easier when doing genetics problems because all of the genotypes have unambiguous phenotypes.

The symbols used for incomplete dominance are sometimes slightly different than for dominant alleles. Lower case letters tend to be reserved for recessive alleles. Often when there is incomplete dominance, each allele will be given a number or a different letter as a superscript. For example R^1, R^2, etc. or I^A, I^B, etc.

Solved Problem 9-1. If you cross true-breeding four-o-clock plants with red flowers to true-breeding four-o-clock plants with white flowers, the resulting heterozygotes have pink flowers. What will be the ratio of red, white and pink flowers if two pink flowered plants are crossed?

Answer Let $R^1 R^1$ be red and $R^2 R^2$ be white. Pink will be $R^1 R^2$. Diagram the cross.

P: $R^1 R^2$ × $R^1 R^2$

Gametes: Half the gametes will be R^1.
Half the gametes will be R^2.

Punnett Square:

	R^1	R^2
R^1	$R^1 R^1$ Red	$R^1 R^2$ Pink
R^2	$R^1 R^2$ Pink	$R^2 R^2$ White

So the ratio will be 1 red: 2 pink: 1 white. The difference from a cross in which red would be dominant is that the pink flowers can be distinguished from the red flowers.

Solved Problem 9-2. A true-breeding radish with long red roots was crossed to a true-breeding radish with round white roots. The F_1 radishes were all oval and purple. Assuming that two independently assorting loci are involved, what would be the ratio of all the possible phenotypes among the F_2 produced by crossing two of the F_1 radishes?

Diagram the cross first:
P: Long Red × Round White

F_1 All Oval Purple

F_2 ??

We know that the parental radishes are all homozygous (true breeding) for two independently assorting loci, probably one for color and one for shape. The F_1 is probably heterozygous for both loci. We know that the F_1 is different from either parental type so it is probably a case of incomplete dominance for both loci. Let's choose some gene symbols:

"Shape locus" S^L = long
S^C = round

"Color locus" C^R = red
C^W = white

(Because no alleles are dominant, we don't use the upper case and lower case symbols.) Now diagram the cross with all that we know.

P: Long Red × Round White
$S^L S^L \; C^R C^R$ $S^C S^C \; C^W C^W$

F_1 All Oval Purple
$S^L S^C \; C^R C^W$

F_2 ??

Now we are ready to answer the question. Cross two F_1 plants, determine the kinds of gametes they would produce, and construct a Punnett Square.

$S^L S^C \; C^R C^W$ will produce four kinds of gametes, $S^L C^R$, $S^L C^W$, $S^C C^R$ and $S^C C^W$. So the Punnett Square will be

	$S^L C^R$	$S^L C^W$	$S^C C^R$	$S^C C^W$
$S^L C^R$	$S^L S^L \; C^R C^R$ Long Red	$S^L S^L \; C^R C^W$ Long Purple	$S^L S^C \; C^R C^R$ Oval Red	$S^L S^C \; C^R C^W$ Oval Purple
$S^L C^W$	$S^L S^L \; C^R C^W$ Long Purple	$S^L S^L \; C^W C^W$ Long white	$S^L S^C \; C^R C^W$ Oval Purple	$S^L S^C \; C^W C^W$ Oval White
$S^C C^R$	$S^L S^C \; C^R C^R$ Oval Red	$S^L S^C \; C^R C^W$ Oval Purple	$S^C S^C \; C^R C^R$ Round Red	$S^C S^C \; C^R C^W$ Round Purple
$S^C C^W$	$S^L S^C \; C^R C^W$ Oval Purple	$S^L S^C \; C^W C^W$ Oval White	$S^C S^C \; C^R C^W$ Round Purple	$S^C S^C \; C^W C^W$ Round White

The phenotypic ratios can be found by counting up the number of each phenotype in the Punnett Square:

Long Red	1/16
Long Purple	2/16
Long White	1/16
Oval Red	2/16
Oval Purple	4/16
Oval White	2/16
Round Red	1/16
Round Purple	2/16
Round White	1/16

PRACTICE PROBLEMS

9-3. Would the results of Problem 9-2 have been any different if we had started with true-breeding radishes that were long white and round red?

9-4. In cattle the coat can be either red $C^R C^R$, white $C^W C^W$ or roan $C^R C^W$. The horns can be either normal (hh) or missing (HH or Hh). The coat color alleles show incomplete dominance, and the horn alleles show dominance. Suppose we crossed two $C^R C^W$ Hh cattle. What proportion of their offspring would be roan and have horns? What proportion would lack horns and be white?

2. Multiple alleles

So far we have limited ourselves to loci with only two alleles, winged vs. wingless, striped vs. solid, disease vs. no disease. But there can be more than two alternate forms of one locus. A single locus represented by 3 or more alleles in a population is said to have **multiple alleles**. Solving problems with multiple alleles is not that much more difficult than solving simpler problems since diploid organisms can still only have two alleles at any locus. The important thing is to be systematic. As long as you consider what is happening in only one individual at a time, every multiple allele problem becomes a collection of problems involving two alleles.

Solved Problem 9-5. It was first thought that the albino locus in rabbits consisted of only two alleles, the wild type C and recessive c. Recessive homozygotes lack melanin and have white hair and pink eyes. Evidence has shown that several other alleles of this locus exist. A slightly more complicated symbol system is used that adds a superscript to alleles. In addition to the dominant C and recessive c allele designations, the "chinchilla" c^{ch} and Himalayan c^h are used to unambiguously designate the multiple alleles for this locus. Any single rabbit inherits only two of these alleles, which segregate in a normal fashion during meiosis. Write the genotypes of the progeny that will result from each of the following crosses.

A) Cc × $c^h c^h$ B) Cc × $c^h c^{ch}$ C) cc × $c^h c^{ch}$

Answer

A) P: Cc × $c^h c^h$

Gametes: ½ C ; ½ c   all c^h

Punnett Square:

	C	c
c^h	C c^h	c c^h

B) P: Cc × $c^h c^{ch}$

Gametes: ½ C ; ½ c   ½ cʰ ; ½ c^ch

Punnett Square

	C	c
cʰ	C cʰ	c cʰ
c^ch	C c^ch	c c^ch

C) P: cc × cʰ c^ch

Gametes: all c   ½ cʰ ; ½ c^ch

Punnett Square

	c
cʰ	c cʰ
c^ch	c c^ch

In problem 9-5 we only looked at the genotypes. When considering phenotypes in multiple allele problems, dominance is a little more complicated. "Dominant" and "recessive" are relative terms that describe what happens in a heterozygote—if there is dominance, the heterozygote will look just like the dominant homozygote. When there are multiple alleles, an allele may be dominant to one allele and recessive to another or show incomplete dominance with one allele and complete dominance with another. Multiple allele problems are much easier if you first get as much genotype information figured out as you can and then work on the phenotypes.

Solved Problem 9-6. The human ABO blood group locus has three alleles, I^A, I^B, and i^o. I^A and I^B are both dominant to i^o, but they show no dominance in I^A I^B heterozygotes. The ABO alleles control antigens on the surface of red blood cells which can be recognized with the appropriate antibodies. Persons whose cells have A antigen are said to be Type A, those with B antigen are Type B, those with both the A and B antigens are type AB, and those without either antigen are Type O. Here is a chart of the relationship of genotypes to phenotypes.

Blood Type	Genotypes
A	$I^A I^A$ or $I^A i^o$
B	$I^B I^B$ or $I^B i^o$
AB	$I^A I^B$
O	$i^o i^o$

What will be the phenotypes produced among the offspring of two people who are group AB? What are the phenotypes of the offspring of a type A heterozygote and a type B heterozygote?

Answer First lay out what is known about genotypes, work that out for parents and offspring and then deal with phenotypes.

Type AB people have the unambiguous genotype $I^A I^B$. The cross looks like this and is like any other cross we've considered with alleles having incomplete dominance.

P: $I^A I^B$ × $I^A I^B$

Gametes: ½ (I^A); ½ (I^B) ½ (I^A); ½ (I^B)

Punnett Square (Fill in the genotypes and <u>then</u> worry about the phenotypes.)

First do genotypes

	I^A	I^B
I^A	$I^A I^A$	$I^A I^B$
I^B	$I^A I^B$	$I^B I^B$

Then add phenotypes

	I^A	I^B
I^A	$I^A I^A$ Type A	$I^A I^B$ Type AB
I^B	$I^A I^B$ Type AB	$I^B I^B$ Type B

In the second cross, all three alleles are involved, but always only two are in any one individual.

P: $I^A i^o$ × $I^B i^o$

Gametes: ½ (I^A); ½ (i^o) ½ (I^B); ½ (i^o)

Punnett Square (Fill in the genotypes and <u>then</u> worry about the phenotypes.)

First do genotypes

	I^A	i^o
I^B	$I^A I^B$	$I^B i^o$
i^o	$I^A i^o$	$i^o i^o$

Then add phenotypes

	I^A	i^o
I^B	$I^A I^B$ Type AB	$I^B i^o$ Type B
i^o	$I^A i^o$ Type A	$i^o i^o$ Type O

Solved Problem 9-7. Mrs. Idengaku was one of two mothers in a maternity ward. When she was given Baby #1, she denied that it was hers, claiming Baby #2 instead. Another mother also claimed Baby #2.

Mrs. Idengaku is Blood type O. Baby #1 is A, and Baby #2 is O. Unfortunately, Mr. Idengaku died just before the baby was born, so we can't find out his phenotype or genotype, but the Idengakus have three other children whose phenotypes are known. Keiko is A, Tohru is B and Kenichi is B. Who is right, Mrs. Idengaku or the other mother? What is your reasoning?

Answer Set up the cross between Mr. and Mrs. Idengaku and put in all the information we have from the problem.

Mrs. Idengaku		Mr. Idengaku
Type O		Unknown blood type
i⁰ i⁰		_ _
Keiko	Tohru	Kenichi
Type A	Type B	Type B
Iᴬ _	Iᴮ _	Iᴮ _
Baby #1	Baby #2	
Type A	Type O	
Iᴬ _	i⁰ i⁰	

There are Idengaku children with both Iᴬ and Iᴮ alleles. Since Mrs. Idengaku is i⁰ i⁰, those alleles must have both come from the father. So Mr. Idengaku was Iᴬ Iᴮ. If so, the Idengakus cannot have a child that is i⁰ i⁰, since Mr. Idengaku had no i⁰ allele to give the baby. They could, however, have a child that is type A. Their type A children would be Iᴬ i⁰. Thus Mrs. Idengaku is mistaken—Baby #2 could not be an Idengaku. Baby #1 could be, however.

PRACTICE PROBLEMS

9-8. With regard to the ABO blood group, one parent in a family is type A and the other is type B. If all four blood types appear among the offspring, what proportion of the offspring will be type O?

9-9. The rabbit coat color genes described in problem 9-5 show a complex form of step-wise dominance. When the c (albino) allele is homozygous, it prevents pigment formation so the homozygous individual has white hair. Allele c^h produces "Himalayan" rabbits with light bodies and dark noses. It is dominant to albino. The c^{ch} (chinchilla) allele produces a coat color that is lighter than wild type but darker than Himalayan, and it produces a "light gray" coat color when heterozygous with c or c^h. Allele C is dominant to all three of the other alleles. What are all the types of coat color that will be seen in the parents and offspring of a cross between a C c^h heterozygote and a c c^{ch} heterozygote?

3. Gene interactions: epistasis

Sometimes alleles at one locus will have an effect on the expression of alleles at another locus. When a gene at one locus masks the expression of genes at another locus, the process is called **epistasis.** The result of epistasis is a modification of the phenotypic ratios you would expect if there were no epistasis. The alleles still assort independently, but some of the phenotypes may be reduced in number or missing. You do these problems exactly as you would any other problems involving two or more genes. When you come to determining the phenotypes, however, you have to be more careful since you have to take those interactions into account.

Solved Problem 9-10. Black mice and brown mice differ in how the pigment melanin is distributed in the fur. Allele B is dominant and results in black fur, while allele b is recessive and results in brown fur when homozygous. Another gene determines whether the pigment melanin is synthesized at all. The recessive c allele blocks melanin production so the resulting "albino" mice have white hair. The dominant C prevents melanin production. Obviously, if there is no melanin production due to homozygosity for c, it does not matter what allele is at the B/b locus. So allele c is epistatic to the B/b locus when c is homozygous. What would be the phenotypes and genotypes of parents and their offspring if we crossed two mice heterozygous for both loci?

Answer Diagram what we know already from the problem.

Bb Cc × Bb Cc

These mice have a C allele so they can make melanin, and they have a B allele so they are black. Each parent can produce four kinds of gametes BC, Bc, bC and bc. The Punnett Square will look like this:

	BC	Bc	bC	bc
BC	BBCC Black	BBCc Black	BbCC Black	BbCc Black
Bc	BBCc Black	BBcc albino	BbCc Black	Bbcc albino
bC	BbCC Black	BbCc Black	bb CC Brown	bb Cc Brown
bc	BbCc Black	Bbcc albino	bbCc Brown	bbcc albino

If this cross of two heterozygotes had involved two loci, each with dominance but with no epistasis, we know that we should expect a 9:3:3:1 ratio of phenotypes among the progeny. But epistasis by the albino allele sometimes masks the color gene. So:

B _ C _	Black	9/16
B _ cc	white	3/16
bb C _	brown	3/16
bb cc	white	1/16

That is 9 black: 3 brown: 4 white. We see the same gene arrangement as usual, but different interactions that produce the phenotypes. The easiest way to work out these problems is first to work out the segregation and independent assortment of genes and then to determine the phenotypes.

Solved Problem 9-11. A gray horse was crossed to a chestnut horse. All the offspring were gray. Then two of the F_1 were crossed. Among the F_2, 12/16 were gray, 3/16 were bay, and 1/16 were chestnut. Explain these results by showing the genotypes of each phenotype among the parents, the F_1 and the F_2.

Answer This problem is the reverse of problem 9-10. This time you are being given phenotype data from which you have to figure out genotypes. In problem 9-10, you were given the genotypes and how the genes interacted and were asked to figure out genotypes. The secret of this second kind of problem is to find a ratio you recognize and build on that.

Diagram what you know:
P: Gray   ×   Chestnut

F_1 all Gray

F_2 12/16 Gray
                3/16 Bay
                1/16 Chestnut

Whenever you see sixteenths, you should think of the 9/16: 3/16: 3/16: 1/16 ratio that results from crossing two organisms that are heterozygous for two independently assorting loci. In this case the 9:3:3:1 ratio appears to have been modified by some sort of epistasis. One of the 3/16 classes has been combined with the 9/16.

F_1		Aa Bb	×	Aa Bb
F_2		9/16 A_ B_		
		3/16 A_ bb		
		3/16 aa B_		
		1/16 aa bb		

The 1/16 category is unambiguously the double homozygous recessive. So chestnut has that genotype. It looks as if the A allele masks the expression of the B and b alleles, so that whenever there is an A allele the phenotype is gray.

F_1		Aa Bb	×	Aa Bb	
F_2		9/16 A_ B_		gray	
		3/16 A_ bb		gray	
		3/16 aa B_		bay	
		1/16 aa bb		chestnut	

And then, of course, the B_ genotype produces the bay phenotype. In problem 9-10, the epistatic allele (albino) was recessive. Here the epistatic allele is dominant (gray) so we get a different modification of the 9:3:3:1 ratio.

One parent was chestnut, so it must have been aa bb. To produce the double heterozygous F_1, the other parent must have been AA BB. We would expect the phenotype to be gray, and it is!

PRACTICE PROBLEMS

9-12. Because of the medical importance of mosquitoes, their genetics has been studied carefully, and many different genes have been found. In one species of mosquito, their larvae are ordinarily tan with a light colored mid-dorsal stripe running along the abdomen. The allele s eliminates the stripe when homozygous. The S allele is dominant and results in the stripe. Another locus controls the body color, with the dominant allele B resulting in tan and the recessive b resulting in dark brown when homozygous. Allele b also reduces the mid-dorsal stripe to a tiny dot on the third segment of the abdomen. Suppose a homozygous normal mosquito is crossed to a dark brown mosquito with neither stripe nor spot. What will be the phenotypes of the F_1 and F_2 generations?

9-13. A black mouse was crossed to a hairless mouse. Since it lacked hair, it was impossible to tell if its hair was black or brown! All the progeny had black hair. When these F_1 progeny were mated among themselves, 100 F_2 were black, 43 were hairless and 30 were brown. What were the genotypes of the original parents?

B. Sex linkage.

Some organisms have a **chromosomal method of sex determination** in which the two sexes have differences in one pair of chromosomes. In humans and fruit flies, females and males differ in a single pair of chromosomes called the **sex chromosomes**. In the females the two sex chromosomes are both **X-chromosomes**, while in the males there is an X-chromosome and a **Y-chromosome**. Often the Y-chromosome contains few genes while the X-chromosome contains many genes, so there are loci on the X that are not on the Y. The genes on the X that are not on the Y are called **X-linked genes**. (Sometimes they are called "sex-linked" genes, but strictly speaking sex linkage includes rare cases of Y-linked genes.) X-linked genes add a complication to genetics problems. All the other chromosomes that are not sex chromosomes are called **autosomes**. All the problems in this book up to this point have involved genes on the autosomes, called **autosomal genes**.

1. Use informative symbols

Problems involving X-linked genes are not difficult to do if you remember that the X and Y chromosomes segregate much as any pair of alleles would segregate and if you use a system of symbols to remind you of sex linkage. A simple way to do this is to write an X or a Y for the chromosome and then write the alleles being studied as if connected to the X. For example:

A female fly heterozygous for the white eye gene (w):	$X^W X^w$
A male fly with the white eye gene:	$X^w Y$
A female human heterozygous for muscular dystrophy:	$X^D X^d$
A male human with the allele for red-green color blindness:	$X^c Y$

Notice that females can be either heterozygous or homozygous, have only one copy of the gene, so those terms can't be applied. Instead we say that males are **hemizygous**.

Solved Problem 9-14. Define symbols and write the genotypes of each of the following individuals.

A) A female fruit fly heterozygous for the X-linked gene for forked bristles

B) A female human heterozygous for the hemophilia gene.

C) A female human homozygous for the hemophilia gene.

D) A male human with the Duchenne's muscular dystrophy gene.

E) A male human without the gene for red-green color blindness.

Answer First write the sex chromosome composition of each individual and then hook on the appropriate allele symbol.

A) A female so XX. Let f be forked bristles. $X^F X^f$

B) A female so XX. Let H be wild type and h be hemophilia. $X^H X^h$

C) A female so XX. $X^h X^h$

D) A male so XY. Let the allele for Duchenne's muscular dystrophy be d. $X^d Y$

E) A male so XY. Let c be the allele for color blindness and C its normal allele. $X^C Y$

2. Keep the alleles attached to their chromosomes.

The X and Y chromosomes segregate into their gametes much as a pair of alleles would. Half the male gametes will receive a Y chromosome and half will receive an X. All the female gametes will receive an X. If the female is heterozygous for an X-linked gene, half of her gametes will contain an X with one of the alleles on it, and half will contain an X with the other allele on it. Remember that, and you can't go wrong.

Solved Problem 9-15. The first allele discovered by Drosophila geneticists was the recessive white-eye allele w, which is X-linked. The normal eye is red. A red-eyed male was crossed to a white-eyed female, and a white-eyed male was crossed to a red-eyed female. All parents are from true breeding lines. What kinds of progeny will be found in each case?

Answer Define symbols and diagram each cross. Let X^W be the symbol for the red-eyed allele and X^w for the white-eyed gene. Remember that the Y has neither allele, so it is just Y.

P:	Red male $X^W Y$	×	White female $X^w X^w$		White male $X^w Y$	×	Red female $X^W X^W$
Gametes:	½ X^W ½ Y		all X^w		½ X^w ½ Y		all X^W

Punnett Squares

	X^W	Y
X^w	$X^W X^w$ Red female	$X^w Y$ White male

	X^w	Y
X^W	$X^W X^w$ Red female	$X^W Y$ Red male

Solved Problem 9-16. In Drosophila forked bristles (f) is an X-linked character. Straight bristles (F) is dominant. Vestigial wings (v) is autosomal, and its wild type allele (V) is dominant. If a male with normal bristles and homozygous for normal wings is crossed to a vestigial winged female that is heterozygous at the forked locus, show the genotypes and phenotypes of their offspring.

Answer Outline the cross.

A) P: VV $X^F Y$ ×   vv $X^F X^f$

         Male   Female
         Normal Wings   Vestigial Wings
         Normal Bristles   Normal Bristles

Gametes   ½ (V X^F); ½ (X^F Y)   ½ (v X^F); ½ (v X^f)

Punnett Square:

	V X^F	VY
v X^F	Vv $X^F X^F$ Female Normal wings Normal bristles	Vv $X^F Y$ Female Normal wings Normal bristles
v X^f	Vv $X^F X^f$ Female Normal wings Normal bristles	Vv $X^f Y$ Female Normal wings Forked bristles

All the females and half the males have normal wings and normal bristles. Half the males have normal wings and forked bristles.

PRACTICE PROBLEMS

9-17. Red-green color blindness is a recessive X-linked character in humans. In cases involving two parents with normal color vision, can male offspring be color blind? Can female offspring?

9-18. Look at the results of Problem 9-16. If one of the male offspring with forked bristles is crossed to a female heterozygous for both loci, how many of each possible type would you expect if there were 40 progeny from that cross?

3. You can sometimes recognize the patterns of sex-linked inheritance.

As with the autosomes, some problems tell you about the parents and ask you about the offspring, but some problems give you information about the offspring and expect you to recognize that the pattern of inheritance reveals sex-linkage. Several clues suggest sex-linkage. Not all of these clues will be found in every case, but any of them should cause you to consider sex-linkage as a possibility.

Clue 1. Among the progeny, the male phenotypes may be in different ratios than the female progeny. Look, for example at the cross in Problem 9-16. The bristle gene is X-linked, and the vestigial gene is not. Both the male and female progeny have normal wings, but some of the male progeny have forked bristles while all of the female progeny have normal bristles. On the other hand, notice Practice Problem 9-18. The males and females have the same phenotype for both wing and bristles. The presence of one of these clues is good evidence for sex-linkage, but the absence of clues does not eliminate the possibility of X-linkage. Sometimes you have to look at several different crosses to be sure.

Clue 2. **Reciprocal crosses** may give different results. In reciprocal crosses, the phenotype of the male in one cross is the phenotype of the female in the other. Problem 9-15 involves reciprocal crosses of the white eye gene. Notice that in one cross all the offspring have red eyes, but in the reciprocal cross the males have white eyes.

Solved Problem 9-19. Suppose a recessive allele l causes death in the early embryo so that homozygotes for l are never counted among the progeny. What would be the effect if the gene were autosomal and a female heterozygous for l were crossed to a normal homozygous male? What if the gene were X-linked and a heterozygous female were crossed to a normal male?

Answer Outline the two crosses.

	Autosomal		X-linked	
P:	LL ×	Ll	$X^L Y$ ×	$X^L X^l$
	Male	Female	Male	Female
Gametes:	all L	½ L ; ½ l	½ X^L ; ½ Y	½ X^L ; ½ X^l
F$_1$	½ LL; ½ Ll		¼ $X^L X^L$; ¼ $X^L X^l$	Females all normal
	All normal		¼ $X^L Y$; ¼ $X^l Y$	Half the males missing
	No difference in the two sexes			

So when the gene is X-linked, there will be half as many male progeny as female progeny. There is no effect on the sex ratio when the gene is autosomal.

Solved Problem 9-20. Red-green color blindness is a defect in the color sensitivity of the retina. It was noticed that it tended to appear in males more frequently than in females and that a male with red-green color blindness could have two normal parents. Usually men with red-green color blindness do not have color blind children, but they can have color blind grandchildren. What kinds of children and grandchildren would you expect in a rare case of a woman with red green color blindness assuming that she married a man without the condition?

Answer That two normal people can have a child with red-green color blindness is evidence that the gene is recessive. That red-green color blind men would have grandchildren with the problem but not children suggests sex-linkage.

Father is X^cY. All his daughters will get the X^c, but his sons will get the Y. If the mother is normal, none of their sons or daughters will be color blind. But grandsons that receive the X^c will be.

If red-green color blindness is recessive and a woman has it, she must be homozygous. She is $X^c X^c$.

Among her children, all the males will get an X^c and thus will be color blind. All her female children will get the normal X^C from the father and thus will not be color blind.

Her sons are X^cY. If they have children by normal wives, none of the grandchildren will be color blind since all get an X^C from their mother. All her daughters, however, will be $X^C X^c$, so half her grandsons will be color blind.

Practice Problems

9-21. The Bar eye allele in <u>Drosophila</u> results in a reduction in the number of units in the compound eye so that the eye is long and narrow instead of round. This allele is dominant to the allele for normally shaped eyes. When males from a true-breeding Bar-eyed strain are crossed to normal females, the female offspring have Bar eyes, but the males do not. When normal males are crossed to females from a true-breeding bar-eyed strain, all the progeny have Bar eyes. Explain this result.

9-22. Here is a cross involving three different independently assorting loci. One of the loci is X-linked. Which one?

P: Long Purple Active Male   ×   Short Yellow Sluggish Female

F_1 1/4 Long Yellow Active Males
           1/4 Short Yellow Active Males
           1/4 Long Purple Active Females
           1/4 Short Purple Active Females

ANSWERS TO PRACTICE PROBLEMS

1-2. P: brown × white

   F_1 all brown

   F_2 1 white, the others brown

1-3. P: yellow marked × white plain

F_1	yellow marked	yellow plain	white marked	white plain
	65	56	61	59

2-3. If the Huntington's locus is H for disease and h for normal and cystic fibrosis is C for normal and c for cystic fibrosis, then

   A) Hh   B) hh CC or hh Cc   C) hh cc

2-4. P: EE × ee

   F_1 all Ee

3-2. Black sheep and spotted goats were true breeding, while solid colored goats and white sheep were not.

3-5. Yes, we can know the genotypes of the true breeding strains since "true breeding" means "homozygous." If we let the symbol B stand for the allele for white sheep and b stand for the allele for black sheep, then Jake's black sheep were bb and the white sheep were B _ (where the blank means tha the other allele could be either B or b). Likewise, if S is the allele for solid colors, and s is the allele for spotted, Jake's spotted goats are ss, and the solid goats are S _.

3-8. Since the flies have the dominant phenotype, they must either be WW or Ww. We can know for certain that they have at least one W allele. So we could symbolize their genotype as W _, with the blank standing for either W or w.

3-9. The runt litter mates are rr, which means that each parent was Rr. Since Honey is not a runt, he got an R allele from one parent, but he could have received either an R or an r from the other parent. So he is either RR or Rr.

3-10. Cross Honey to a runt pig (rr). If Honey is RR, all the offspring will get an R from Honey and will therefore all be normal. But if Honey is Rr, some offspring will get an r from Honey. They must get an r from the runt, so some of Honey's offspring will be rr and runts.

4-2. All true breeding (homozygous) so if their genotype is WW, all the gametes will be W.

4-3. Both parents are heterozygous, so each will produce A and a gametes in equal proportions.

4-8. Both parents must be heterozygous for cleft palate. Let P be normal and p be the cleft palate allele. Each pig can produce both P and p gametes.

4-9.

EFTV	eFTV
EFTv	eFTv
EFtV	eFtV
EFtv	eFtv
EfTV	efTV
EfTv	efTv
EftV	eftV
Eftv	eftv

5-3. Each individual produces only one kind of gamete, so the Punnet Square will be very simple:

	M N
m n	

5-4. Bb Ff Gg individuals will produce 8 kinds of gametes. So there will be 8 gametes on each side of the square.

	BFG	BFg	BfG	Bfg	bFG	bFg	bfG	bfg
BFG								
BFg								
BfG								
Bfg								
bFG								
bFg								
bfG								
bfg								

6-2.

	M N
m n	Mm Nn

6-3. This one will have 8 × 8 = 64 intersections!

	BFG	BFg	BfG	Bfg	bFG	bFg	bfG	bfg
BFG	BB FF GG	BB FF Gg	BB Ff GG	BB Ff Gg	Bb FF GG	Bb FF Gg	Bb Ff GG	Bb Ff Gg
BFg	BB FF Gg	BB FF gg	BB Ff Gg	BB Ff gg	Bb FF Gg	Bb FF gg	Bb Ff Gg	Bb Ff gg
BfG	BB Ff GG	BB Ff Gg	BB ff GG	BB ff Gg	Bb Ff GG	Bb Ff Gg	Bb ff GG	Bb ff Gg
Bfg	BB Ff Gg	BB Ff gg	BB ff Gg	BB ff gg	Bb Ff Gg	Bb Ff gg	Bb ff Gg	Bb ff gg
bFG	Bb FF GG	Bb FF Gg	Bb Ff GG	Bb Ff Gg	bb FF GG	bb FF Gg	bb Ff GG	bb Ff Gg
bFg	Bb FF Gg	Bb FF gg	Bb Ff Gg	Bb Ff Gg	bb FF Gg	bb FF gg	bb Ff Gg	bb Ff gg
bfG	Bb Ff GG	Bb Ff Gg	Bb ff GG	Bb ff Gg	bb Ff GG	bb Ff Gg	bb ff GG	bb ff Gg
bfg	Bb Ff Gg	Bb Ff gg	Bb ff Gg	Bb ff gg	bb Ff Gg	bb Ff gg	bb ff Gg	bb ff gg

7-3. For any one allele, a cross of two heterozygotes will produce homozygous recessives 1/4 of the time. So the chance of being homozygous for recessive alleles at 10 different loci is $1/4 \times 1/4 \times 1/4 \times 1/4 \times 1/4 \times 1/4 \times 1/4 \times 1/4 \times 1/4 \times 1/4$ or $(1/4)^{10}$, which is 0.00000095 or about 1 in one million. (Think if you had to do this with a Punnett Square!)

7-4. Diagram the cross, including all the information in the problem:

P: Rr hh   ×   rr Hh
         Black and white   Red and white
         Horned cow   Hornless bull

Gametes: ½ (Rh); ½ (rh)   ½ (rH); ½ (rh)

Punnett Square

	Rh	rh
rH	Rr Hh black/white hornless	rr Hh red/white hornless
rh	Rr hh black/white horned	rr hh red/white horned

Red and white hornless occurs 1/4 of the time. Assuming that males and females occur in equal numbers, we should expect 1/8 of the offspring to be red and white hornless cows.

There are 2 × 15 = 30 offspring. So 1/8 of 30 or about 3 or 4 offspring will be red and white hornless cows.

The father is a red and white hornless bull. Red and white hornless also occurs 1/4 of the time, so those that look like the father will occur 1/8 of the time and again about 3 or 4 of the 30 progeny will be like that.

8-3. In the cross of Ff Gg Hh × ff gg hh, the triple heterozygote will produce 8 different gamete types, and the triple homozygote will produce only one kind (f g h). So each of the possible offspring will receive one of the eight possible gametes form the heterozygote and one f g h gamete. All of the possible offspring are listed below.

Ff Gg Hh	ff Gg Hh
Ff Gg hh	ff Gg hh
Ff gg Hh	ff gg Hh
Ff gg hh	ff gg hh

8-4. Let's let D be the allele for polydactyly and d its recessive while p is the allele for phenylketonuria and P is the allele for lack of phenylketonuria. Since the man's father had phenylketonuria, the father must have been pp, so the man is Pp. The man doesn't have the dominant polydactyly either, so he is dd. The woman has polydactyly, so she is Dd. Since the problem specifies that she has no phenylketonuria allele, she is also PP. Here is a diagram of what we know about this cross:

P: Pp dd   ×   PP Dd

Gametes: ½ (Pd) ; ½ (pd)   ½ (PD) ; ½ (Pd)

Punnett Square:

	Pd	pd
PD	PP Dd polydactyly	Pp Dd polydactyly
Pd	PP dd neither condition	Pp dd neither condition

There is no chance that any offspring will have both conditions, nor is there a chance that any offspring will have phenylketonuria. Half the offspring can be expected to have polydactyly.

8-7. The ratio is nearly 9:3:3:1, which is characteristic of a cross of two parents heterozygous for two independently assorting genes. So the parents were
               Dd Nn   ×   Dd Nn

8-8. Write what we know of the genotypes of Izzy and Lucinda.

Izzy	Lucinda
bb ww F_	B_ W_ ff

If Lucinda's other parent had big white spots, weak hooves and at least one f allele (B_ W_ _f), then it is possible that Izzy could be Lucinda's parent.

9-3. The results will be the same since the F_1 radishes will still be heterozygous for both loci.

9-4. We can use the Product Rule on this one. Consider the cross as two separate crosses — $C^R C^W \times C^R C^W$ and $Hh \times Hh$.

The proportions that have horns and are roan is $\frac{1}{4}$ (those with horns) $\times \frac{1}{2}$ (those that are roan). $\frac{1}{4} \times \frac{1}{2} = \frac{1}{8}$.

The proportion that lack horns and are white is $\frac{3}{4} \times \frac{1}{4} = \frac{3}{16}$.

9-8. For an $i^o i^o$ child (type A) to appear among the offspring, each parent must contribute an i^o allele. For AB to appear among the offspring, one parent must contribute an I^A and one must contribute an I^B. So one parent must be $I^A i^o$ and one must be $I^B i^o$. The cross is thus the same as in Problem 9-7, and type O offspring will occur $\frac{1}{4}$ of the time.

9-9. P: $C\ c^{ch}$ $\times$ $c\ c^h$
               Normal   Himalayan

Gametes: $\frac{1}{2}\ \text{\textcircled{C}};\ \frac{1}{2}\ \text{\textcircled{c^{ch}}}$ $\frac{1}{2}\ \text{\textcircled{c}};\ \frac{1}{2}\ \text{\textcircled{c^h}}$

Punnett Square:

	C	c^{ch}
c	Cc Normal	$c\ c^{ch}$ light gray
c^h	$C\ c^h$ Normal	$c^{ch}\ c^h$ light gray

9-12. The problem involves a gene that is not quite completely epistatic. Allele b suppresses, but does not eliminate, the stripe produced by allele S.

F_1	All Bb Ss	tan with a stripe	
F_2	B_ S_	tan with a stripe	9/16
	B_ ss	tan with no stripe or spot	3/16
	bb S_	brown with a spot	3/16
	bb ss	brown with no stripe or spot	1/16

Since epistasis is not complete, it's the same old double heterozygote problem with a 9:3:3:1 ratio. This emphasizes that the segregation and assortment of genes is not changed by epistasis. The only new issue is remembering how the genes interact to produce phenotypes.

9-13. Diagram what you know from how the problem is stated.

P: black $\times$ hairless

F_1 All black

F_2 Black   100
               Hairless 43
               Brown   30

The F_2 ratio is close to a 9:4:3 which is a modified 9:3:3:1. So the F_1 must be double heterozygotes. Let

34   *A Problem-Based Guide to Basic Genetics*

$\underline{B}$ be black and $\underline{b}$ brown. Let $\underline{H}$ be haired and $\underline{h}$ be hairless. Then the F_1 and F_2 are:

F_1 All black   Bb Hh

F_2 Black   100   B_ H_   9/16
       Hairless 43   B_ hh 3/16 and bb hh 1/16
       Brown   30   bb H_ 3/16

The parents must have been Black BB HH and Hairless bb hh.

9-17. If the mother is heterozygous, she will have normal color vision. Half of her offspring will receive her X-chromosome which includes the color-blind gene. Each male that receives that X will receive a Y chromosome from the father and so will be color blind. Each female that receives the color blind gene from the mother will receive a normal X from the father, so they will not be color blind.

9-18.

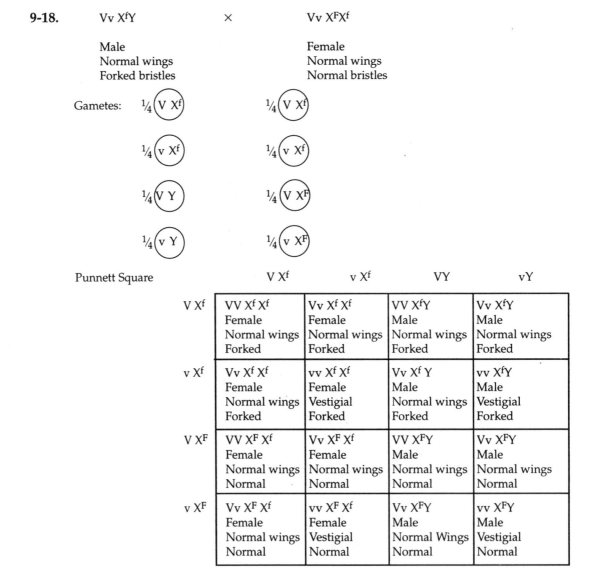

Among the males, 6 have normal wings and 2 have vestigial for a 3:1 ratio. Also, 4 have normal bristles and 4 have forked bristles for a ratio of 1:1.

Among the females, there is also a 3:1 ratio of winged to vestigial and a 1:1 ratio of forked to normal bristles.

Of 400 flies, 200 will be females, and of those 3/8 or 75 will have both normal wings and normal bristles, 3/8 or 75 will have normal wings and forked bristles, 1/8 or 25 will have vestigial wings and normal bristles and 1/8 or 25 will have vestigial wings and forked bristles.

Similarly the males will include 75 normal normal, 75 normal forked, 25 vestigial normal and 25 vestigial forked.

9-21. The difference in reciprocal crosses suggests X-linkage. Let $\underline{B}$ symbolize the Bar allele and $\underline{b}$ symbolize the normal. Put the alleles on the X-chromosome and the results are explained.

P: $X^b X^b$ × $X^B Y$ $X^B X^B$ × $X^b Y$
    Female   Male   Female   Male
    Normal   Bar   Bar   Normal

F_1 Males all $X^b Y$ (normal)   Males $X^B Y$ (bar)
        Females $X^B X^b$ (bar)   Females $X^B X^b$ (bar)

9-22. Look at the characters one at a time.

1. Long X short gives a ratio of 1:1 in a 1:1 sex ratio. We can therefore assume that this locus is not X-linked and that one parent is a heterozygote (Ll) and one is a homozygote recessive (ll).

2. Active X sluggish results in all active progeny regardless of sex. It is a result of a mating of two homozygous parents where active (A) is dominant over sluggish (a). AA X aa.

3. Purple males mated with yellow females results in yellow males and purple females. Color segregates by sex, so we can deduce that the color locus is X-linked. Purple (X^P) is X-linked dominant over yellow (X^p). $X^p X^p$ X $X^P Y$